ORCA FOOTPRINTS

What Poo Can Do

HOW ANIMALS ARE FIGHTING THE CLIMATE CRISIS

YOLANDA RIDGE

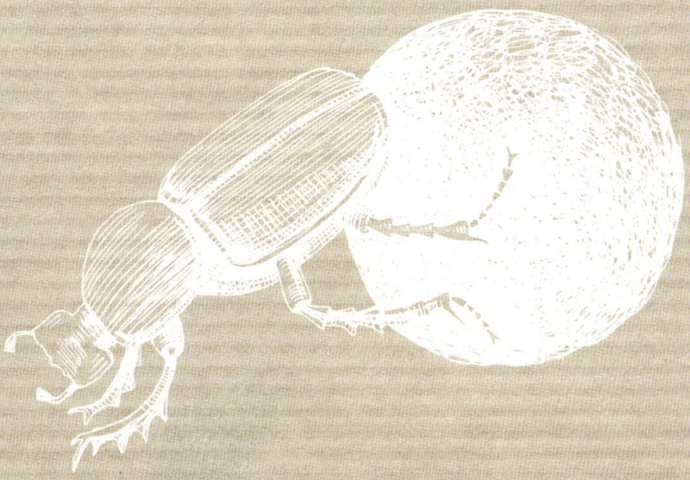

ORCA BOOK PUBLISHERS

Published in Canada and the United States in 2024 by Orca Book Publishers.
Also available as an ebook (ISBN 9781459835429, PDF;
ISBN 9781459835436, EPUB).
orcabook.com

Library and Archives Canada Cataloguing in Publication
Title: What poo can do : how animals are fighting
the climate crisis / Yolanda Ridge.
Names: Ridge, Yolanda, 1973- author.
Series: Orca footprints ; 29.
Description: Series statement: Orca footprints ; 29 |
Includes bibliographical references and index.
Identifiers: Canadiana (print) 20230229441 | Canadiana
(ebook) 2023022945X | ISBN 9781459835412 (hardcover) |
ISBN 9781459835429 (PDF) | ISBN 9781459835436 (EPUB)
Subjects: LCSH: Animal droppings—Juvenile literature. |
LCSH: Climatic changes—Juvenile literature. |
LCSH: Animal ecology—Juvenile literature. |
LCSH: Climate change mitigation—Juvenile literature. |
LCSH: Animal behavior—Juvenile literature.
Classification: LCC QC903.15 .R53 2024 | DDC j363.738/74—dc23

Library of Congress Control Number: 2023934912

Summary: This nonfiction book for middle-grade readers, illustrated with photographs throughout, explores how animals are fighting the climate crisis by pooping.

Orca Book Publishers is committed to reducing the consumption of nonrenewable resources in the production of our books. We make every effort to use materials that support a sustainable future.

Orca Book Publishers gratefully acknowledges the support for its publishing programs provided by the following agencies: the Government of Canada, the Canada Council for the Arts and the Province of British Columbia through the BC Arts Council and the Book Publishing Tax Credit.

Front cover photos by jrpolofoto/shutterstock.com
and Paul A. Souders/Getty Images
Back cover photos by vladsilver/Getty Images,
arnolgs/pixabay.com and Efimova Anna/Shutterstock.com
Design by Jenn Playford
Edited by Kirstie Hudson

Printed and bound in South Korea.

27 26 25 24 • 1 2 3 4

Wildebeests see a lot of other animals during their seasonal migration, including pink flamingos, which gather in big groups along rivers and lakeshores.
MASSIMO LAMA / EYEEM/GETTY IMAGES

For Tim. Thank you for smiling every time I say "poo"—
and for everything else too.

Contents

CHAPTER THREE
DUNG BEETLES

CHAPTER FOUR
PENGUIN GUANO

Whales poop near the surface of the water when they come up to breathe.
CHASE DEKKER WILD-LIFE IMAGES/GETTY IMAGES

Introduction

Whales do something you do too—they poo!

Did you know that animals like whales and wildebeests help fight ***climate change*** every time they take a dump? It's true! And that's not all. The poop left behind by creatures like dung beetles and penguins provides clues about how the ***climate crisis*** is affecting the ***ecosystem***.

In this book you'll find out more about the climate crisis and why it matters—not just to us but to every living creature on Earth. We'll explore how specific animals are saving the planet by fertilizing plants, taking ***greenhouse gases*** out of the air, preventing forest fires and more…all through the power of poo!

By finding ways to help certain animals, you can join the fight against the climate crisis too. A big part of saving Earth from rapid climate change is protecting the animals we share it with. Not just one animal or a couple of animals but all of them—including ourselves.

The Climate Crisis

Even though it's normal for Earth's climate to change slowly over time, it's currently changing about 10 times faster than ever before. This rapid change is related to human activity such as

- burning ***fossil fuels*** like coal, oil and gas to make electricity and fuel our cars and planes;
- cutting down forests to make room for houses, factories and farming;
- raising more cattle and growing more food.

Our planet can only handle so much change at once. Right now things are changing so fast that many people consider it a climate crisis.

CHAPTER ONE

Whale Poop

Blue whales like to snack on krill. They swim into swarms of krill with their mouths open wide! APPLE PHO/SHUTTERSTOCK.COM

Let's start this "poop-tastic" journey in the ocean. Why? Because even though you might think of climate change as something that happens to the air, changes in ocean temperature have a huge impact. More on that later, but for now follow me into the depths of the sea, where a blue whale's getting ready for lunch.

A WHALE OF A MEAL

You're about to find out what it's like to dine with the largest animal in the world. Most fully grown blue whales are over 486 tons (440 metric tons)—the combined weight of three space shuttles!—and more than 100 feet (30 meters) long. Given its massive size, Blue's lunch is different than you might think. Instead of chowing down on steak or loading up on carbs, the blue whale's diet consists almost entirely of krill—a small shrimp-like creature about the size of your thumb.

Whales hold their breath when they dive deep into the water to eat.
RICHCAREY/GETTY IMAGES

How does a whale get to be so big eating a creature so small? By consuming as many as 40 million of them per day! Luckily for the whale, krill hang out near the bottom of the ocean in large, dense masses commonly referred to as clouds or swarms. To satisfy their hunger, blue whales swim into one of these krill swarms with their mouths open, engulfing both water and prey. Once it's got a mouth full, the blue whale filters its catch by pushing the water out with its tongue through ***baleen*** bristles, which are a bit like the teeth of a comb. The water goes out and the krill stay in (sometimes joined by fish, other ***crustaceans*** and even the odd squid). The thing about blue whales is that they can't eat and breathe at the same time. So after they're finished gobbling up krill, they jet up to the surface of the ocean to get some air.

Weather vs. Climate

Weather describes what's happening outside your window. When you say it's raining, you're talking about the weather. *Climate* refers to the general weather conditions expected in a specific area of the world at a particular time of year. In Canada, the climate is colder in the winter than it is in the summer.

OCEAN FERTILIZER

Even though whales can do a number two wherever they want—the ocean is literally one big toilet for them—research shows that whales most often relieve themselves near the surface. (People who study whale poo describe it as "fecal plumes with richly odiferous qualities." In other words, it stinks!)

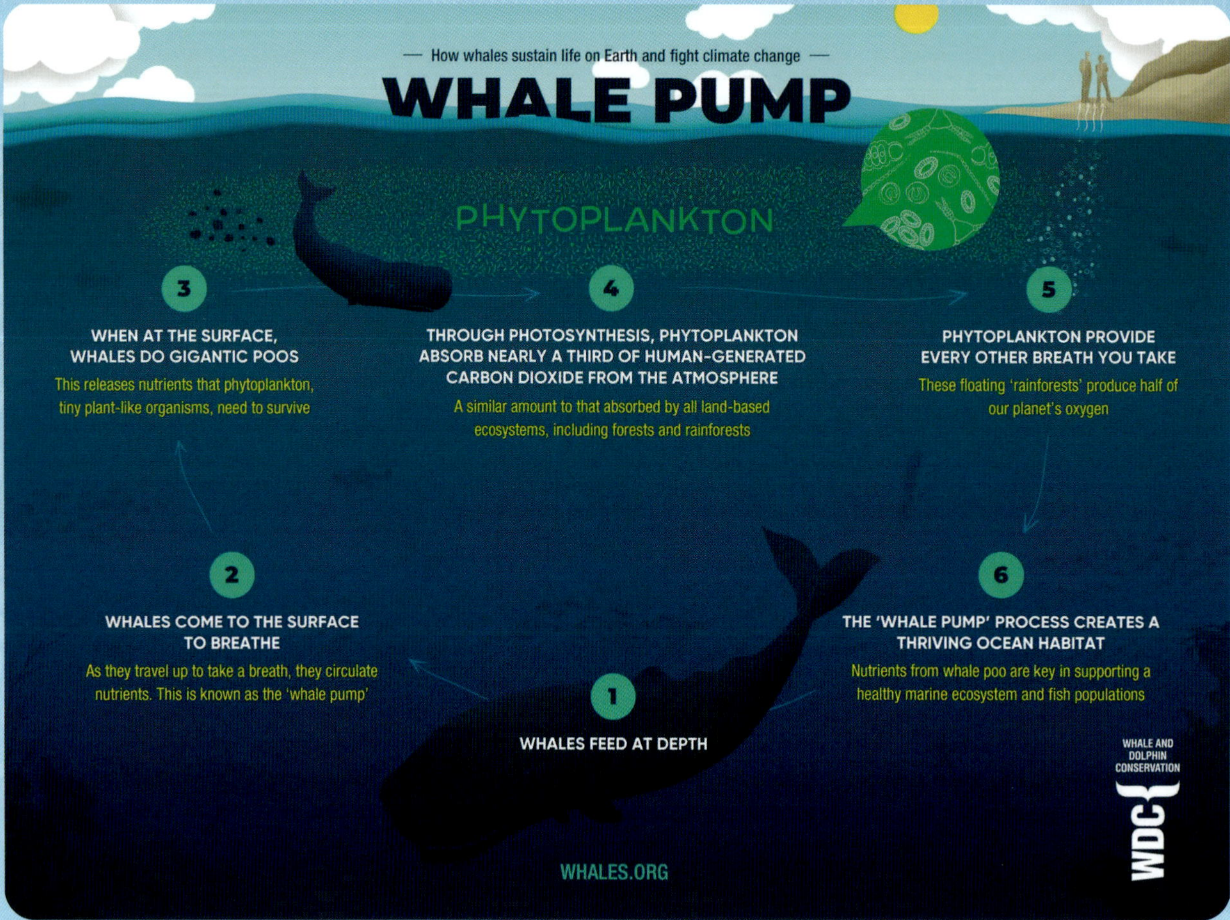

Blue whales can dive to a depth of 1,640 feet (500 meters), but they usually feed about 330 feet (100 meters) under the water. When they come up for air, they move—or pump—nutrients to the surface.
COURTESY OF WHALE AND DOLPHIN CONSERVATION, WHALES.ORG

Before whale bones get washed onto shore, the carcass sinks to the bottom of the ocean. One whale fall can feed and protect an entire community of deep-sea creatures for many years. IVSTIV/GETTY IMAGES

Why does this matter? Because it means that the whale is bringing nutrients—in the form of digested krill—from the bottom of the ocean to the top. This creates what scientists refer to as a "whale pump" (although "poop pump" sounds better).

The whale's nutrient-rich clouds of poop float through the surface water, supporting the growth of phytoplankton—a microscopic plant that absorbs ***carbon dioxide*** and produces ***oxygen*** through ***photosynthesis***. In the same way that compost fertilizes the tomatoes growing in your garden, whale poop fertilizes phytoplankton growing in the ocean.

What does all this have to do with climate change? Carbon dioxide is one of the main greenhouse gases contributing to the climate crisis. Like all plants, when phytoplankton gets energy through photosynthesis, it takes carbon dioxide out of the air and replaces it with the oxygen most animals need to breathe. More phytoplankton equals less carbon dioxide, and less carbon dioxide means less climate change.

KEEPING CARBON DOWN

In addition to taking carbon dioxide out of the air by feeding phytoplankton with their poo, blue whales store carbon in their bodies. As you can imagine, a creature the size of a space shuttle can store a lot of carbon. When a blue whale dies, it takes all this carbon to the bottom of the ocean floor. This prevents the carbon from being released into the atmosphere and adding to the climate crisis. The whale's body also provides habitat and food for many deep-sea creatures that rely entirely on whale ***carcasses*** to survive.

Photosynthesis

Have you ever wondered what plants eat when they get hungry? It's sunshine! The cells that make up a plant's leaves, stems and flowers all absorb sunlight. Inside some of these cells, a special ingredient called *chlorophyll* traps the light so it can be used for photosynthesis. During photosynthesis, hydrogen molecules from water and carbon dioxide from the air mix together to make a sugary snack for the plant. The oxygen left over from this process is then released into the air.

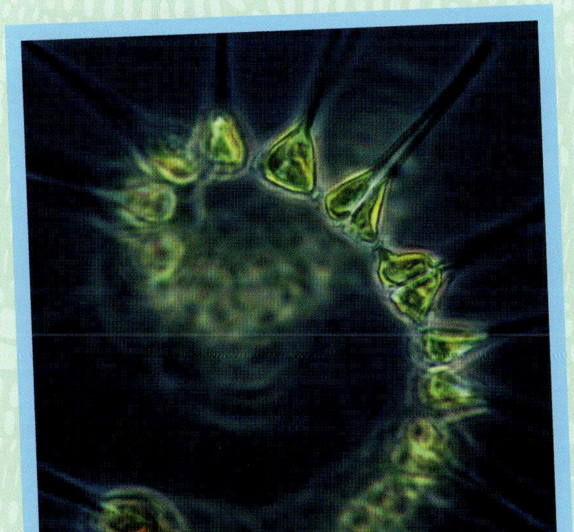

FOTOSHOPTOFS/PIXABAY.COM

Phytoplankton

Nine out of ten plants that live in the ocean are considered phytoplankton, including the seaweed and algae you might have seen (or swam into) on a trip to the beach. Phytoplankton is considered the base of the food chain because it's eaten by so many small fish, who get eaten by bigger fish, who get eaten by even bigger fish... you get the idea. At least half the oxygen in the air we breathe is produced by phytoplankton. In the process of putting oxygen into the air, phytoplankton takes a lot of carbon dioxide out of the atmosphere through photosynthesis—a lot of work for a tiny plant!

Scientists believe humpback whales breach to communicate with other whales. When their fins and flukes slap the water's surface, the sound can be heard by whales that are very far away. HECKEPICS/GETTY IMAGES

The Magical Element for Everything on Earth

All life on Earth depends on carbon—including you! In fact, you're made of it. If you weigh 100 pounds (45 kilograms), you contain a whopping 18 pounds (8 kilograms) of carbon. Carbon is also found in things like rocks, fossil fuels, plastic products and even your pencil. It's the fourth most common element in the universe and combines with other elements in the periodic table to make compounds. Combine one carbon atom with two oxygen atoms and voilà—you've got carbon dioxide.

HOW CAN WE HELP WHALES FIGHT THE CLIMATE CRISIS?

Here's what you shouldn't do: poop in the ocean. Not only is your doo-doo useless for the phytoplankton, but untreated sewage can cause disease in ocean animals and land animals (including us!). Here's what we can do: support efforts to stop whaling and reduce plastic waste.

Whaling

People have hunted whales for thousands of years. Indigenous Peoples from different parts of the world, such as the Inuit (Arctic), Basque (Europe) and Ainu (Japan), have relied on whales so much for survival that whaling has become part of their cultural identity.

In these cultures, nearly every part of the whale is used. Everything that can be eaten is eaten. Meat, blubber and even the organs provide important protein, fat, vitamins and minerals. (Imagine how many people can be fed with a tongue as big as a hippo!) Parts that can't be consumed, like baleen bristles, are woven into baskets or used to make fishing lines. Even the bones are used to make tools or ceremonial items like masks.

Today most whaling is done for reasons other than survival, often by people outside these traditional cultures. Things started to change in the late 1800s, for two main reasons:

- Whales became easier to hunt because of technology such as gun-loaded harpoons and steamships (that might not sound high-tech, but it was at the time).

- People started hunting whales for the oil found in whale fat, which was used to light lamps before there was electricity.

Because of this, whale hunting became a multimillion-dollar industry, and some scientists estimate that more whales were hunted in the early 1900s than in the previous four centuries combined.

Close to Extinct

Things changed again when people started using fossil fuels as a source of energy and stopped using whale oil. Unfortunately, it happened too late to save the whales. Many whale species—including our friend Blue—were already close to becoming ***extinct***. And you know what that means: no whales, no whale poo.

In 1986 the International Whaling Commission tried to suspend commercial whaling through what's called a moratorium. Not all countries agreed to stop hunting whales, but many did. Since then whale populations have started to recover. Fifty years after whaling wiped out all the Antarctic blue whales living around the island of South Georgia, 58 were spotted there in 2020.

Many individuals and groups organize protests against whale and dolphin hunting to raise awareness and send a message to policymakers. LOREDANA SANGIULIANO/ SHUTTERSTOCK.COM

How You Can Help

Now it's up to us to continue making things better for the whales. Here's how you can get involved:

- Join an organization that's working to protect endangered whale species, such as the World Wildlife Fund (WWF).
- Write a letter to your local politicians asking them to stop whaling. If your country's already part of the International Whaling Commission you can ask them to put pressure on the countries that still have commercial whaling industries, such as Norway, Japan and Iceland.

Reduce Plastic Waste

Today the biggest threat to whales isn't whalers. It's destruction of the whale's habitat. For whales to keep fighting the climate crisis, they need a healthy ocean to eat, reproduce, play and poop in.

Many things pollute our oceans, but plastic is one of the worst. At least 9 million tons (8.1 million metric tons) of plastic is dumped into the ocean every year—more than a garbage truck load every minute! Where is this plastic coming from and how does it get into the ocean? Among the top 10 kinds of trash picked up during the 2021 International Coastal Cleanup were food wrappers, beverage bottles, grocery bags, straws and takeout containers. These things—and many others—make their way into our waterways from boats, garbage that's not properly disposed of and litter that gets blown, swept or washed out to sea. Unlike other trash, plastic doesn't decompose. Once it gets into the ocean, plastic waste either floats or sinks. All of it gets tossed around, and a lot breaks into tiny pieces called microplastics.

Single-use plastic water bottles do not decompose, and they remain in the ocean for hundreds of years, along with other plastic pollution. MR.ANAKED/SHUTTERSTOCK.COM

Plastic Impact

Plastic pollution is responsible for the death of many ***seabirds*** and marine animals, including whales. There are two main ways plastic kills: entanglement and ingestion. Entanglement mainly affects smaller animals. Having plastic wrapped around a flipper or fin can prevent a sea creature from moving. Plastic trash can also wrap itself around mouths, noses and blowholes, making it hard for animals to eat and breathe. For a whale as large as Blue, the main risk comes from fishing gear such as ropes and nets.

Ingestion is exactly what it sounds like—eating plastic. Whales have been found washed up on shore with bellies full of everything from flip-flops to big tangles of nylon rope. Since the blue whale filters food through baleen bristles, it's less likely than other marine ***mammals*** to consume large pieces of plastic. But microplastics can get through Blue's comblike teeth.

With a belly full of plastic, whales can't get the food or water they need to survive. Microplastics can also build up inside an animal's tissues, where they leach toxic chemicals that make the animal sick. This doesn't just happen to Blue. It also happens to Blue's prey, which means the krill the whales had for lunch could also make them sick.

Trash that enters the ocean in one place often gets washed to shore somewhere completely different—sometimes even another country. Volunteers around the world are tackling this problem through coastal cleanups.
SOLSTOCK/GETTY IMAGES

How You Can Help

Unlike the situation with whaling, plastic pollution is only getting worse. Here are a few ways you can help:

- Reduce your own plastic consumption, not just through recycling but by using less. There are lots of great ways to get started, and some of them are fun, like getting ice cream in a cone so you don't use a plastic spoon or bowl. You can also skip the straw, fill your reusable water bottle at the fountain, bring reusable bags to the store, pack your lunch in reusable containers and avoid using face wash and toothpaste containing microbeads.

- If you live near the ocean, organize a coastal cleanup or join one already in progress. For information on how to do this, check out websites such as Ocean Conservancy's Fighting for Trash Free Seas or Ocean Wise's Great Canadian Shoreline Cleanup.

- Join an organization like Ocean Alliance, which is working to make the ocean environment safe for whales through research and education on plastic, chemical and noise pollution.

CHAPTER TWO

Wildebeest Droppings

We're on safari, traveling from the bottom of the ocean to the plains of Africa. Only in this case, we're not here just to watch animals. We're joining them on a long walk to see how wildebeests—and the droppings they leave behind—fight the climate crisis with every step.

Blue wildebeest calves must be able to migrate with their herd shortly after birth.
KENCANNING/GETTY IMAGES

SEASONAL MIGRATION

Every year millions of wildebeests migrate in search of greener pastures. For blue wildebeests (which, like blue whales, are not actually blue) the trek begins in the southern Serengeti—a large ecosystem in Tanzania that has protected areas and game parks.

The exact timing of their departure depends on the weather. They leave at the end of the rainy season—usually late May—when food and water in their grassland home become scarce.

Wildebeests do most of their walking in the early morning and late afternoon. They take breaks when it gets too hot. WAYNE MARINOVICH/GETTY IMAGES

During their trek they walk up to 1,000 miles (1,600 kilometers). It would take us more than 10 days to walk that far without any breaks even to sleep!

For wildebeests, the trek takes much longer—about seven months. They take the same path from south to north every year, gathering around the Mara River—the only permanent source of water in the area—during the dry season. In late October, when rain returns to the southern grasslands, they head home to give birth to their calves. They don't go back the way they came, though. Instead the wildebeests always follow the same circular route through the Serengeti. Along the way, they are active day and night, grazing constantly. Their favorite thing to eat is short grasses, which cover the southern plains in the wet season. Which brings us back to the poop they leave behind!

Gnu I Know You?

What's the difference between a wildebeest and a gnu? Nothing! In Afrikaans (a language spoken in South Africa), the gnu is called *wildebeest*—"wild beast"—because of its large head and menacing horns.

It's not just wildebeests that leave number two behind as they migrate. Giraffes, zebras and gazelles do it too!
R. MAXIMILIANE/SHUTTERSTOCK.COM

Animals like lions, cheetahs, wild dogs and hyenas rely on wildebeest meat for food. Wildebeests come together when threatened by predators, stamping and giving out alarm calls to drive away the threat.
JAMES WARWICK/GETTY IMAGE

SEED SPREADER

Wildebeest manure fertilizes the grasses and trees along the animals' migratory path—just like whale poop fertilizes phytoplankton. All those droppings do something else too: they distribute seeds so more plants can grow. Spreading seeds might not seem like a big deal. But this seasonal migration isn't just a big deal—it's huge! As many as 1.5 million blue wildebeests participate in the Serengeti migration every year. They're joined by many other animals, including zebras and gazelles, which have a different diet but still produce lots of poop.

These animals all migrate to stay alive. It's the only way they can keep their bellies full through the entire year. If they stayed in one place, many would die. (Quite a few wildebeests die during migration too, during risky maneuvers like crossing rivers full of crocodiles, but that's another story.)

Gardeners of the Grassland

Unlike animals, plants can't just walk to the next buffet when they run out of food and drink. Instead they must rely on wildebeests to deliver nutrient-rich dung so they can stay alive when times are tough. And when the going gets too tough, they count on wildebeests to eat their seeds and poop them out in different places—hopefully, somewhere they're more likely to grow. Wildebeest gardening duties don't end there. They also help the soil by moving it around with their noses and mouths as they graze. This ***aerates*** the land and makes it better for growing—a bit like what happens when we dig in the garden and turn dirt over with a shovel.

What does all this have to do with the climate crisis? Like phytoplankton, trees and grass on the Serengeti plains breathe in carbon dioxide and breathe out oxygen. The trees—although small compared to some in North America—also store carbon in their trunks, keeping it out of the atmosphere so it can't go back to being a greenhouse gas.

Acacia trees grow on the African plains. Long, sharp thorns can be hidden among the leaves, so giraffes have to be careful when reaching for a nibble with their long tongues.

Poopy Compost

Most of us think of compost as rotten banana peels, apple cores and other food waste. But anything containing carbon—even poop—can be broken down by ***microorganisms*** to create soil that has nutrients, such as nitrogen, phosphorus and potassium (elements that help plants grow).

Wildebeests like to graze on grass. And the grass benefits from the nutrients provided by wildebeest dung. COOLVER/SHUTTERSTOCK.COM

FIREFIGHTERS

So what happens when a tree—or a whole forest—burns down? Carbon is released into the air, where it combines with oxygen to create carbon dioxide again. This is another way wildebeests fight climate change—by joining the fire department! Not really, of course. They don't arrive in a fire truck to spray the flames with water. And they can't stop fires from starting due to natural events like lightning or from human-related causes like driving hot vehicles over dry grass.

But they do stop fires from spreading in two ways:

- Wildebeests make the soil rich with their grazing and pooping and keep the plants healthy by nibbling away both old wood and new shoots—like we do when we prune a tree to make it stronger. An ecosystem that contains strong, healthy plants is much less likely to burn out of control.

- When wildebeests walk and graze along the same path every year, they leave a bare track of land behind. This creates what's called a ***firebreak*** because there's nothing left to burn. A fire that starts at one end of the Serengeti can't reach the other end if it's stopped by a firebreak.

HOW CAN WE HELP WILDEBEESTS FIGHT THE CLIMATE CRISIS?

There are lots of ways we can help the wildebeest. Going on safari isn't one of them, unfortunately, although ***ecotourism*** can help local people in places like the Serengeti by giving them financial security and a reason to preserve their natural resources. The best way to help wildebeests fight the climate crisis is to become seed spreaders and firefighters, just like them. We can do this by rewilding and by stopping deforestation.

Rewilding

The wildebeest population, unlike that of blue whales, is currently stable and not considered at risk thanks to rewilding—a process of returning nature back to its wild state by reducing, removing or reversing the impact of human activity. To understand why rewilding was necessary, we have to go back about a hundred years, to a time when a virus spread from farm animals caused the Serengeti wildebeest population to plummet. With fewer wildebeests around, the stuff they normally eat grew out of control on the African plains. Lots of grass and trees might sound like a good thing—all that carbon being sucked out of the air—but remember that vegetation must be healthy to be helpful. Plus, with a smaller number of migrating wildebeests came smaller firebreaks that weren't as effective at preventing fires from spreading. Eventually, having fewer wildebeests in the Serengeti led to massive wildfires that destroyed up to 80 percent of the ecosystem. Not just once—every single year. The amount of carbon dioxide released into the atmosphere from burning trees and grass turned the Serengeti into a place that produced more greenhouse gas than it absorbed.

A firebreak can help control wildfires by stopping or slowing down their progress.
THOMMY TFH/SHUTTERSTOCK.COM

Serengeti Sink

People took action to stop the spread of disease from farm animals to wild animals in the 1960s. Within 10 to 20 years, the number of wildebeests returned to historic levels and brought the landscape back into balance. With wildebeests grazing and migrating across the plains once again, wildfire decreased. And the Serengeti went back to absorbing more greenhouse gas than it produced. Today the area drains as much carbon from the atmosphere as East Africa puts into it through fossil fuel emissions, earning it the nickname Serengeti Sink.

The result was profound, and many people consider it one of the most successful rewilding stories ever. Scientists have shown that bringing large mammals back to an area where they used to live can also reduce the risk of flooding, droughts, ***invasive species***

Greenhouse Gases

When smoke rises from a chimney or puffs out of a car's exhaust pipe, most of us think it's gone for good. It's not. The greenhouse gases in that smoke and exhaust end up blanketing Earth and stopping heat from escaping. They're called *greenhouse gases* because they turn the earth into a greenhouse—warm air comes in through the glass but can't get out. The top two greenhouse gases affecting our planet are carbon dioxide and methane.

Controlled Burns

Not all fires are bad. A controlled burn is a fire that's started on purpose for a specific reason, usually to prevent future forest fires. Some fires—like the ones that burn out of control—are not so good. Wildfires started by accident are especially dangerous to people, animals and the environment. As the climate changes, these types of fires are happening more often. And every time a living thing burns, carbon is released into the air, which adds to the climate crisis.

With moisture, oxygen and warmth, seeds have all they need to grow into seedlings that can be planted into the ground.
RAWPIXEL.COM/GETTY IMAGES

and ***biodiversity*** decline. In addition to these things—which will all become more common with climate change—rewilding can help fight the climate crisis.

Rewilding efforts—and success stories—aren't limited to the Serengeti or even Africa (although they do have lots of big mammals there). Rewilding is happening in the Arctic, where reindeer (and their poop) could help stop ***permafrost***—which stores a lot of carbon—from melting. The same concept's also being used in rainforests to protect and reintroduce ***keystone species*** like gorillas, tapirs and toucans, which are especially important to the health of big trees.

How You Can Help

The term rewilding describes any conservation effort focused on letting nature take care of itself. Even though it has a lot to do with limiting human involvement—or interference—there are many ways you can support it:

- Celebrate World Rewilding Day on March 20 by posting about it online and getting involved in activities organized by Rewilding: The Global Alliance.
- Visit the Rewilding Earth website to listen to podcasts, read poems and learn about programs aimed at everything from protecting jaguars to developing wildlife corridors.
- Rewild your own space—a backyard garden or a pot on the balcony—by planting flowers that would grow there if you weren't around. This will help attract butterflies, bumblebees and other critters that were part of the ecosystem before we altered it.

Naturally occurring fires can be beneficial to both plants and animals. But when ecosystems aren't healthy, wildfires can burn out of control and become dangerous.
JAMES WARWICK/GETTY IMAGES

Gardeners of the Forest

Like wildebeests, gorillas are so important to the health of jungle trees and plants that they're often called gardeners of the forest. They keep tropical forests (which store 40 percent of the world's carbon) healthy by fertilizing plants with their poop and removing brush that could cause fires. Gorillas eat different things depending on the season and whether they live in the mountains or the lowlands. Like all primates, gorillas love fruit but also enjoy leaves, roots, shoots, seeds and even bark.

KIKI DOHMEIER/SHUTTERSTOCK.COM

Deforestation

Letting nature heal itself through rewilding sounds great. It would be better, though, if we prevented nature from becoming damaged in the first place. One example of this is stopping deforestation. Why do forests get cut down? Usually because people want to use the land for something else, like growing food or building houses. As the human population continues to rise, there will be more demand for these spaces. One alternative to cutting down trees is making better use of the land. Instead of new single-family houses, for example, we could build multifamily dwellings where people share common spaces like gardens and garages.

Preserving the Forest

According to the ***United Nations (UN)***, deforestation currently accounts for 17 percent of carbon emissions worldwide. That's one reason the UN developed a program that provides incentives for countries to preserve their forests. Countries that have lost the most forest over the past 10 years tend to be in developing parts of the world such as Brazil, Indonesia and Tanzania. That's why it's important that wealthy countries, through organizations like the UN, work with less wealthy countries to find alternatives to deforestation.

The good news is that deforestation has decreased in recent years. The other thing that's become more common is the planting of new trees when a forest does get burned down. This isn't as effective at fighting the climate crisis because small trees take in less carbon, but it's better than no trees at all.

How You Can Help

Even though a lot of work to stop deforestation needs to be done at the government level, there are ways you can get involved. Here are a few ideas:

- Try not to buy things that require deforestation. One example is palm oil, a product that's in everything from packaged groceries to shampoo. It often comes from palm trees grown on land that's been deforested. Because of how this contributes to the climate crisis, many companies are finding alternatives to palm oil or sourcing it from ecologically safe trees. You can make a difference by choosing products with labels that say "no palm oil" or "RSPO-certified palm oil."

- Plant a tree! Do it in your local community or support the Nature Conservancy's Plant a Billion Trees campaign by asking friends and family to give you the gift of a tree planting for your birthday or other special occasion instead of a present.

Healthy trees in the Amazon rainforest stand next to a field that's been scorched by fire. In many places, trees are often burned down so people can plant crops.

BLOOMBERG CREATIVE/GETTY IMAGES

Every time you plant a tree, you're helping animals fight the climate crisis. Plants that make the biggest difference are often the ones that naturally grow in your area or ecosystem.

CLAUDIAD/GETTY IMAGES

CHAPTER THREE

Dung Beetles

Dung beetles are very strong. The male of one species can roll dung balls that are over a thousand times heavier than it is. That's like an average-sized person moving six school buses!.
HENK BOGAARD/SHUTTERSTOCK.COM

Down on the farm, there's another greenhouse gas contributing to the climate crisis—***methane***. To understand where it's coming from, we need to take a closer look (and smell) at the poo-producing part of the body. That's right, we're going right to the guts of the matter! We'll also dive into the resulting piles of poop to meet a climate-crisis hero—the dung beetle.

BURPING, FARTING AND POOPING IT OUT

Before we dig into the dung, let's get up close and personal with the farm animal that creates the most of it. When a cow eats, the food is digested by microorganisms to get nutrients and energy. As these microorganisms break down food like hay and grass, methane is produced. The gas isn't useful to the cow, so it becomes a by-product of digestion that needs to be, well...released.

When we burp, fart and poop, we're releasing methane too. We just don't release quite so much of it because our stomachs

Watch your tail! Cow bottoms fart out methane and poop out dung.

LYNN_BYSTROM/GETTY IMAGES

are different than cows' and what we eat is different too. A single cow belches out about 220 pounds (100 kilograms) of methane a year—equal to the weight of a newborn elephant. With more than a billion cows on the planet, that's a lot of methane! As anyone who's walked across a cow field can tell you, lots of cows also equals lots of poop. Cow pies are mostly made of methane, which contributes significantly to the climate crisis.

DUNG DESTROYER

Where there's dung there are dung beetles. You may have seen photos of them—these beetles are famous for being able to roll poop into a ball bigger than they are. But dung beetles can do more than one trick. In fact, they do a lot of the dirty work that keeps ecosystems healthy.

Dung beetles are a group of insects that lay eggs and feed off the poop of large mammals. They're found all over the world,

Methane vs. Carbon Dioxide

Greenhouse gases get a bad rap, but we actually need them to maintain life on Earth. Without greenhouse gases in the air, Earth's surface would be too cold. The problem is that human activity is adding too much greenhouse gas to the atmosphere. Worldwide, three-quarters (or 75 percent) of the greenhouse gas added to the air by humans is carbon dioxide. The amount of methane is small in comparison (around 15 percent), but it still has a big impact on climate change. In fact, methane is more than 25 times stronger than carbon dioxide when it comes to trapping heat in the atmosphere.

A shovel-like head and paddle-shaped antennas help dung beetles dig through poo. In some species, the male also has a curved horn on the top of its head. SVSE/PIXABAY.COM

and they come in three main varieties—rollers, tunnelers and dwellers. All types of dung beetle, whether they prefer giraffe manure or cow feces, use poop to raise their babies and put food on the table. But while they're rolling, tunneling or dwelling in dung to find food and shelter, they're doing their part to fight the climate crisis, too, by reducing methane.

How do beetles reduce methane? By making feces fluffy! It might sound gross, but dung beetles aerate poop every time they dig into a pile of it. And cow pats full of air don't release as much methane. To figure out just how much of an effect the dung beetles have, scientists put cow pats with and without beetles into closed chambers and measured how much gas was released. The cow pats with beetles released almost half as much methane as the cow pats that were beetle-free.

Most dung beetles are dark brown or black, and many are shiny. They vary in size from 0.2 to 1.2 inches (0.5 to 3 centimeters), but most adults are about as big as a quarter. WANIDA TUBTAWEE/SHUTTERSTOCK.COM

GROWING VEGETABLES

Preventing this much methane from being released into the air could make a big difference when it comes to the climate crisis. More research needs to be done to show how much impact dung beetles have. It likely depends on such things as the species of dung beetle and where they live.

The good news is that digging into cow pats—or the dung of any other large mammal—does something besides reducing methane. It makes soil healthy and delivers nutrients to plant roots. This means that, like whales and wildebeests, dung beetles help vegetation grow into plants that fight the climate crisis by breathing in carbon dioxide.

HOW CAN WE HELP DUNG BEETLES FIGHT THE CLIMATE CRISIS?

There are a couple of ways we can help dung beetles fight the climate crisis, and neither of them involves walking around with a Save the Beetles sign. We can give them less work to do by changing what we eat and what we feed the cows. We can also support farmers who are changing the way they farm.

Meat versus Plants

As the human population grows, we need more farms to feed us. More farms means more manure and more greenhouse gases. All farm animals add greenhouse gases to the environment by exhaling carbon dioxide and belching or farting out methane. Dairy and beef production accounts for two-thirds (around 60 percent) of all emissions that come from livestock (which includes animals like pigs and chickens).

Plant products, on the other hand, inhale carbon dioxide and exhale oxygen. This alone means that eating fruits and vegetables contributes less to climate change, although there are some nonmeat products (like rice) that can contribute to greenhouse gas emissions depending on how they're grown.

Another thing to consider is that most farm animals eat plants. So when you chow down on beef, you're not just eating cow. In a way, you're also eating the grass that made that cow big and strong (and juicy!). Looking at it this way, it's obvious why producing steak is harder on the environment than making a loaf of bread—you have to grow wheat and other grains to feed the cow (which also needs land and produces lots of methane) instead of just growing the grains.

Growing plants for us to eat is a lot easier on the environment than growing plants to feed livestock. PETER GARRARD BECK/GETTY IMAGES

How Hot Is Too Hot?

Depending on where you live, warmer weather might not seem so bad. But higher temperatures won't turn the world into one big beach resort. It will warm the ocean and change ocean currents, causing parts of Earth to be hotter while others are colder. This means that plants, animals and even people can no longer live where they used to live or eat what they want to eat because of things like flooding and droughts, more severe storms and rising sea levels due to melting polar ice.

Skip the Beef

A report on climate change from the United Nations says that eating less meat will help combat the climate crisis by

- cutting down on greenhouse gas emissions associated with farming livestock;
- reducing the need to create more farmland through deforestation as the human population grows.

Cows and Cud

Cows don't have just one stomach—they have four. And they don't chew their food just once—they chew it twice. Chewing cud (partially digested food that gets returned to the mouth from the stomach) helps cows get more nutrition from less food. It also produces lots of burps and farts!

TONY C FRENCH/GETTY IMAGES

Studies show that if every person in the United States stopped eating meat and dairy for just one day a week, it would reduce greenhouse gas emissions by the same amount as removing 7.6 million cars from the road. The effect would be particularly strong when people in North American and European countries do it. Why? Because people in richer countries tend to eat more meat.

Around the world, 75 Big Macs are sold every single second (which adds up to 6.48 million burgers a day!). Producing the meat for just one of these beef patties creates almost 5 pounds (2.11 kilograms) of carbon dioxide. That number might not mean a lot until you compare it to a plant-based patty made by the same company, which produces less than half a pound (0.12 kilograms). So choosing a veggie burger over a beef burger—even if you do it just once—can make a difference.

How You Can Help

Fortunately, it's getting easier to switch to a more plant-based diet. Here are two ways you can change your diet and give the beetles less work to do:

- Participate in Meatless Monday by eating a meat-free meal once a week. Check out the Meatless Monday website or the cookbooks in the resource section to find recipes.

- Decide what to eat by measuring greenhouse gas emissions per serving. Chicken and cheese, for example, produce less greenhouse gas than beef and lamb. Foodprints calculators, like Eat Low Carbon and Plate Up for the Planet, can help you make these decisions.

Think Before You Chew

Fast-food restaurants are responding to customer demand for meat that's better for the environment. This puts pressure on farmers to care for their land and raise their animals in different ways. To help dung beetles specifically, farmers can use fewer pesticides and other chemicals that destroy insect populations. A study in Ireland showed that cattle farms using artificial fertilizers and insecticides had fewer dung beetles than farms that did not.

The other way farmers can work with dung beetles to reduce emissions is by letting cattle roam and graze on outdoor pastures. When cows are kept in a barn all the time, their dung produces more methane gas because there are fewer beetles inside. There are farmers who believe that farming can be part of the solution to climate change through ***regenerative agriculture***.

Building with Dung

In countries like Kenya, India and Nepal, dung paste is applied to floors, walls and roofs to make homes waterproof and provide insulation. When dung is dried and mixed with mud, it doesn't smell unpleasant. (Mosquitoes don't seem to like the smell, though, which is why dung also acts as an insect repellent.) A group of students from Indonesia took this idea one step further by creating building bricks from cow dung. Compared to clay bricks, dung bricks are both lighter and stronger. They're also better for the environment, easy to make and inexpensive.

Composting is not something you do just at home. It's also one of five methods farmers use in regenerative agriculture.

KILITO CHAN/GETTY IMAGES

Many people have reduced the amount of animal products in their diet in response to the climate crisis—especially young people.
MIXETTO/GETTY IMAGES

In Harmony with Nature

Farmers practice regenerative agriculture by allowing their cattle to graze on one patch of grassland—aerating it as they eat and fertilizing it as they poop (with the help of dung beetles, of course)—for a certain amount of time. After a day or so, the herd moves to a new patch of land. The patch they've just been on then has time to recover, while the vegetation benefits from the air and nutrients the cows have put into the soil.

Another way farmers can more directly reduce methane emissions is by changing what they feed their cattle. Just as our farts, burps and poops smell different depending on what we eat, cows produce different gases depending on their diet. Digesting corn produces less methane than digesting hay and grass. And research shows that adding seaweed to a cow's diet can reduce the methane in its burps by more than 80 percent!

How You Can Help

Back to the beetles, you can help farmers who are working to reduce their cattle's environmental impact with every trip to the shop. Here are some specific ways:

- When you crave a burger with a big glass of milk, choose products labeled organic, pasture-raised, grass-fed or free-range. Each of these terms means a slightly different thing, but they all tell you the cattle have been allowed to graze for food outside the barn. Certain websites can help you find these products.

- Reduce the number of cows in the barn (along with chickens in the coop, pigs in the pen, crops in the field...) by cutting down on food waste. You can do this by menu planning, helping your family make a grocery list before going to the store, modifying recipes so you only make the amount of food you need and freezing anything you can't eat while it's still fresh.

EULALIA CAYUELA MARTINEZ/GETTY IMAGES

There are approximately 1.5 billion cows in the world—that's one cow for every five people! The countries with the most cows are India, Brazil, China and the United States.

STEFANO EMBER/SHUTTERSTOCK.COM

CHAPTER FOUR

Penguin Guano

Baby penguins stay at home on the ice with Dad while Mom travels to the coast to fish. On her way she helps spreads nutrients in the form of penguin poo and partly digested food.
MEMORYCATCHER/PIXABAY.COM

Now it's time to visit a place few people are brave enough to go to—Antarctica. The white patch at the bottom of the globe, Antarctica is the coldest, windiest and driest continent on Earth. Why are we going somewhere with such wicked weather? To meet the emperor.

WHAT A LOAD OF GUANO

Emperor penguins—the largest of all penguins, standing about as tall as a first grader—live nowhere on Earth besides Antarctica. With its freezing temperatures and large expanses of ice, Antarctica might not seem like a haven of biodiversity. But some areas of Antarctica have more life than others, and it turns out that where there's life, there's penguin poo. We've talked a lot about how good poop is for soil. But guano—a fancy word for dried-out bird poop that has accumulated over time—is especially nutritious and full of stuff that helps plants grow, like

You can tell where penguin colonies have huddled together to keep warm by the poop piles left behind.
IGNACIO PALACIOS/GETTY IMAGES

nitrogen, phosphorus and potassium. In fact, guano's so good for soil and plant health that companies make lots of money selling it as fertilizer.

Emperor penguins don't really fly, but that doesn't stop them from spreading their poop all over the place. Like the blue wildebeest, emperor penguins are famous for their yearly migration. They shuffle up to 100 miles (160 kilometers) inland to lay their eggs.

When penguin guano stays where it's dropped, it nourishes the surrounding ecosystem. This includes the rocky outcroppings where small plants like mosses and lichens cling to life in Antarctica. The ocean and its inhabitants—especially the phytoplankton we talked about in chapter 1—also benefit when guano is washed into the sea. Then there's ammonia—a gas that comes out of guano just like methane comes out of dung—which gets picked up by the wind and is blown inland, where it provides nitrogen to the soil there.

Antarctic hair grass, one of only two flowering plants native to Antarctica, grows on rocks covered in guano.
GERALD CORSI/GETTY IMAGES

This photograph was taken by a European Space Agency satellite. The penguins are too small to see, but the patches of guano they produce are very identifiable.

BROWN STREAKS FROM SPACE

In addition to being a great fertilizer, penguin guano does something that might surprise you—it leaves a trail that can be used to study climate change itself. And believe it or not, emperor penguins leave behind enough poop that these trails are visible from space!

As big as emperors are, they're difficult to count because they live in such remote areas. Instead of traveling along the entire coastline of Antarctica and all its islands, scientists search space satellite photos to locate penguin colonies by looking for guano, which shows up as rusty-brown streaks across Antarctica's white ice. Since they need ice, land and access to both water and food, the presence or absence of penguins—and their poop—is a sign of how the climate is changing over time. A place that once had lots of ice, for example, may no longer have enough to support a penguin colony. If this is happening more than it did in the past, it tells us that ice is disappearing faster than before because of climate change caused by humans.

Scientists often visit penguin colonies identified on satellite images to collect information. By digging through mounds of guano, they can find out even more about the penguins' past and present, which helps them predict the future.

Flying Fertilizer Factory

Seabirds are everywhere. They inhabit both poles, the tropics and far-flung islands that humans have never set foot on. Through their travels, nearly all seabirds are excellent nutrient cyclers. Just like the whale brings nutrients from the bottom of the ocean to the top, seabirds bring nutrients from sea to shore when they eat fish and poop out the by-products of digestion as they fly or waddle away.

Adélie penguins get covered in pink poop after sliding around on the ice.
LIAM QUINN/WIKIMEDIA COMMONS/CC BY-SA 2.0

POOP MAPS

Among the things scientists search for when they dig through emperor penguin poop are clues to what they're eating. Adélie penguins, however, allow scientists to get this information without a shovel. How? By the color they leave behind. Like its big cousin, the emperor, the Adélie produces enough excrement to be seen from space, which has helped scientists locate new colonies using satellite imagery. Instead of brown streaks, however, Adélie penguins create "color-coded poop maps."

Shades of Pink

When Adélie penguins have access to an all-you-can-eat krill buffet, their poo takes on a reddish-pink hue. This penguin guano stains the rocky outcroppings where they fish and shows up brightly on the ice where they keep their eggs and babies safe from ocean predators. But when krill are hard to find, Adélie penguins supplement their diet with fish. This makes their poop less pink and almost bluish.

Understanding how the penguins' diet is changing over time provides valuable information on what's happening in the ocean—and how it's being affected by greenhouse gas emissions. As mentioned in chapter 1, climate change isn't something that only affects the air. As greenhouse gases trap more energy

Cloud Creator

Would you believe that seabirds produce enough waste to form clouds? It's true! Ammonia from guano can get sucked into the atmosphere, where it combines with water vapor and sulfuric acid to create low-lying clouds. This plays a role in fighting the climate crisis because sunlight bounces off those clouds and back into space, keeping both air and water cooler—especially in places where there are lots of seabirds like penguins.

from the sun, the ocean absorbs more heat. This is especially true at the surface of the water, where krill go to dine on phytoplankton. But since the water in the ocean is constantly moving, it also affects water at the depth of the ocean where krill hang out to avoid predators like the emperors and Adélies.

Ocean water doesn't just move up and down. It flows around the world as well. This means that Antarctic krill are affected by what happens everywhere on the planet, not just where they live. In some parts of Antarctica, krill populations appear to be declining, in part because of warmer ocean temperatures. But as you can imagine, counting krill is even harder than finding emperors—that's why it's helpful to look at what Adélies are snacking on instead.

HOW CAN WE HELP PENGUINS FIGHT THE CLIMATE CRISIS?

Making sure penguins don't go extinct is the most important thing we can do to help. One activity that puts penguins and other seabirds at risk is fishing. The biggest danger, however, comes from the climate crisis itself. Although it's a threat to all creatures on Earth, emperor penguins—an endangered species—are especially vulnerable to rising temperatures.

Fishing with friends and family is a great way to spend an afternoon. Following rules and guidelines helps ensure enough fish are left to keep the ecosystem healthy.
TERRY VINE/GETTY IMAGES

Commercial Fisheries

Have you ever seen penguins belly flop onto the ice and slide into the water? (If you haven't but want to, there's a video link in the resources section.) The emperors aren't playing—although it does look like fun—but doing a deep dive to snatch a snack.

As good as they are at fishing, emperors run into trouble when there's more than just fish in the sea. We're not talking about their natural predators, like leopard seals and orcas. We're talking about stuff left behind by commercial fishing, like fishing nets and baited hooks. If an unlucky emperor gets tangled up in this kind of fishing gear, it can get dragged behind a fishing boat or sink to the bottom of the ocean. Even though emperors can hold their breath for a long time—up to 12 minutes—they often drown.

Find Better Ways to Fish

Fisheries management that includes guidelines on how, when and where people may fish can prevent this type of tragedy. Rules like these, in addition to limits on how many fish can be caught, also ensure there's enough food left for the penguins. Studies have shown that where intense fishing occurs, seabirds starve.

It's a simple equation. Fewer seabirds equals less guano equals less fertilizer equals less healthy vegetation around the world. Good management can prevent this while still allowing people in the fishing industry to make a living.

Seabirds spend most of their time flying—and pooping—over the water, only coming to land when it's time for a rest.

JAMES WARWICK/GETTY IMAGES

How You Can Help

Having rules isn't helpful, though, if they're not enforced. Here are a few specific ways you can help:

- Follow the rules yourself. Even if you're not heading out to sea in a fishing boat, many small lakes have guidelines—such as what type of fish can be caught and when—to keep fish populations healthy.
- Support commercial fisheries that follow the guidelines by making sure the fish and seafood that make it onto your plate were caught in a way that supports ocean conservation. Organizations like Ocean Wise make this easy with product labels and a searchable website.

Life without Ice

In 2016 one large emperor colony went extinct in Antarctica's Halley Bay after a season with less ice than normal. When the sea ice broke up early, 10,000 baby penguins drowned because they weren't old enough to have the waterproof feathers adults do.

The Ocean Roller Coaster

Even when the ocean's surface appears calm, there's a roller coaster underneath. Warm water from the top joins cold water from the bottom for an epic ride from one end of the earth to the other. As the water rises, falls, spins, circulates and bumps into things, it cycles heat and energy around the planet. These ocean currents affect everything from the daily weather forecast to long-term climate change. And the roller coaster is long—it takes 1,000 years for water to get back to the beginning of the ride!

Climate Change and the Future

The climate crisis is already changing the habitat of many living things, making it harder to get food and survive harsh weather events. It's not just plants and animals that are affected; it's people too. If we don't do our best to end the climate crisis, scientists predict we'll face many problems in the future. There will be more crop failures, and people will lose their homes. Diseases will also become more common, including pandemics like COVID-19, as people and animals come into closer contact in the search for safe, habitable space.

Using your own power to get around is healthy for both you and the planet.
IMGORTHAND/GETTY IMAGES

Scientists are worried this will happen more and more if penguins aren't able to adapt fast enough to the climate crisis on their icy Antarctic habitat. Warm ocean temperatures are already causing ice at both the north and south poles to disappear. It's affecting the health of krill and fish populations as well.

Stop Burning Fossil Fuels

So far we've talked about actions that can help specific creatures fight climate change. But here's the thing. If we don't take responsibility and act as soon as possible, there won't be any penguins left to help. Why? Because penguins need ice—especially the emperors, who rely on it to breed and feed. One study that looked at warming trends and severe weather events related to the climate crisis predicted that 98 percent of emperor penguin colonies will disappear by 2100 if no changes are made to greenhouse gas emissions. That may seem like a long way away, but more than half of those colonies will go extinct by 2050—when you will probably be about the age your parents are now.

How You Can Help

Saving the penguins requires a major change in the way we do things here on Earth. The good news is that we know what we need to do. And together we can do it. Here are the most important steps we can take to fight climate change:

- Stop burning fossil fuels. We hear a lot about how much driving and other forms of transportation contribute to greenhouse gas emissions. The reality is that anything we use that requires energy—including your phone, computer and even the toaster—could be powered by fossil fuels. To cut down on how much energy you use, you can do little things like turning off lights when you don't need them and walking or biking places instead of getting a car ride. You can also do big things, like supporting politicians, companies and environmental organizations that are trying to reduce the use of fossil fuels in the production of goods and transportation.

- Start using ***sustainable energy***. Let's face it. No matter how hard we try, we're not going to stop needing energy to fuel the lifestyles we enjoy. That's why it's important to support initiatives that aim to replace fossil fuels with energy that comes from clean, renewable sources like the sun and wind. Maybe you could even come up with a new source of energy yourself! Some of the best solutions come from "outside-the-box" or creative thinking—something adults aren't always good at doing.

Students build a windmill in science class to learn about alternative ways of creating energy. SDI PRODUCTIONS/GETTY IMAGES

THE POO SOLUTION

From dining in the ocean to going on safari to cleaning up after cows to waddling across the ice, we've come a long way on this fecal journey. We've focused on whales, wildebeests, dung beetles and penguins, but all animals—and plants—make a difference when it comes to climate change.

Poop can't fight the climate crisis alone, of course. But the truth is that healthy ecosystems rely on poo-producing animals and the plants that benefit from their excrement. A big part of protecting our planet is protecting the animals we share it with by reducing human activities that put them at risk, including climate change itself.

It's up to us to limit the greenhouse gas emissions that are causing the climate to change more rapidly than ever before. Humans created the climate crisis. Since we're the most powerful creature on Earth, it's up to us to end it.

Together we can fight the climate crisis...one poop at a time.

Cows eat mostly grass, which farmers also refer to as silage *or* hay. *As a treat, cows enjoy fruit, like apples and bananas.*
BLEND IMAGES - NED FRISK/GETTY IMAGES

Glossary

aerates—circulates air through something such as soil

baleen—fine, comb-like plates found in the mouths of some whale species instead of teeth, used to filter food out of water

biodiversity—the variety of animal and plant species found in a particular ecosystem or environment

carbon dioxide—a gas with no color or smell that's produced by respiration and by burning carbon

carcasses—the dead bodies of animals

climate change—shifts in temperatures and weather patterns over a long period of time

climate crisis—the global threat posed by rapid changes to Earth's climate system, which are currently causing an overall rise in temperature and increased extreme weather events across the planet

crustaceans—arthropods (animals with a body made of segments, a tough outer shell, two pairs of antennae, and limbs that are jointed) that mostly live in the water, such as crabs, lobster and shrimp

ecosystem—a complex system in which everything that exists in a particular environment, including plants, animals, rocks, weather, soil, water, sunlight, etc., relies on the other parts of that environment in the same way

ecotourism—a form of tourism involving responsible travel to natural areas, conserving the environment and improving the well-being of the local people

extinct—a species that has totally died out and can no longer be found anywhere in the wild or in captivity

firebreak—an area without vegetation that creates a break in the landscape to stop fires; created by natural features like rivers, animals like wildebeests and people like foresters

fossil fuels—the remains of dead plants and animals that are extracted from below Earth's crust and burned to create energy

greenhouse gases—gases in Earth's atmosphere that trap heat

invasive species—any species of plant, animal, fungus or bacterium that causes harm when introduced to a new habitat, where it has no natural enemies and therefore can reproduce and spread unchecked

keystone species—a species on which other species largely depend or that helps to keep the shared habitat in balance

mammals—animals that have a backbone or spine, are warm-blooded and feed their babies milk

methane—a gas with no color or smell found in nature and produced by human activity

microorganisms—organisms that can only be seen through a microscope, including bacteria, algae, yeast, mold and viruses

oxygen—a gas with no color or smell that makes up 21 percent of the air on Earth

permafrost—a layer of soil that remains frozen throughout the year

photosynthesis—the process by which plants use sunlight water and carbon dioxide to create oxygen and energy in the form of sugar

regenerative agriculture—a way of farming and ranching in harmony with nature

seabirds—birds that live most of their lives near the ocean

sustainable energy—energy that can be used again and again to meet the needs of the present without putting future generations at risk

United Nations (UN)—an international organization that brings countries together to maintain peace and security

Resources

Print

Drake, Jane, and Ann Love. *Rewilding: Giving Nature a Second Chance.* Annick Press, 2017.

Gleeson, Erin. *The Forest Feast for Kids: Colorful Vegetarian Recipes That Are Simple to Make.* Abrams Books for Young Readers, 2016.

Klein, Naomi, and Rebecca Stefoff. *How to Change Everything: The Young Human's Guide to Protecting the Planet and Each Other.* Atheneum Books for Young Readers, 2021.

Mulder, Michelle. *Brilliant! Shining a Light on Sustainable Energy.* Orca Book Publishers, 2013.

Pollan, Tracy, Dana Pollan and Corky Pollan. *Mostly Plants: 101 Delicious Flexitarian Recipes from the Pollan Family.* Harper Wave, 2019.

Tate, Nikki. *If a Tree Falls: The Global Impact of Deforestation.* Orca Book Publishers, 2020.

Online

8 Things to Know about Palm Oil (WWF-UK): wwf.org.uk/updates/8-things-know-about-palm-oil

Earth Day: earthday.org/foodprints-calculators

Eat Low Carbon: eatlowcarbon.org

Eatwild: eatwild.com/products

The Monday Campaigns: mondaycampaigns.org/meatless-monday

National Aeronautics and Space Administration (NASA): climatekids.nasa.gov

National Geographic Kids: kids.nationalgeographic.com/explore/science/climate-change

The Nature Conservancy: nature.org/en-us/get-involved/how-to-help/plant-a-billion

Ocean Conservancy: oceanconservancy.org/trash-free-seas/international-coastal-cleanup

Ocean Wise Seafood: ocean.org/overfishing/sustainable-seafood/

Ocean Wise Shoreline Cleanup: ocean.org/pollution-plastics/shoreline-cleanup/

Plate Up for the Planet: plateupfortheplanet.org

Rewilding Earth: rewilding.org

World Rewilding Day: globalrewilding.earth

The World Wildlife Fund (WWF): worldwildlife.org

Videos

Climate Change: Earth's Giant Game of Tetris—Joss Fong
TED Ed YouTube Channel

Emperor Penguins Speed Launch Out of the Water
National Geographic YouTube Channel

Renewable Energy 101
Student Energy YouTube Channel

What Happens If Earth Gets 2°C Warmer?
Popular Science YouTube Channel

Index

Page numbers in **bold** *indicate an image caption.*

Acknowledgments

Everything worth doing takes time, energy and teamwork. Writing's no exception, and I'm extremely grateful to all those who've stuck with me on my ever-winding path in and out of publishing (and back again).

To my partner of 20 years (plus!)—I hope my dedication makes you smile. Thank you for helping me become all the things that matter to me—wife, mother, author—and for being with me on the journey.

To my sons—watching you both grow up has been the biggest joy of my life. In a way, you've watched me grow as well, not just as a mother but as an author and person. Thank you for your unconditional love and support.

To the rest of my family—no matter how often we connect, you're always in my heart. Thank you for building my foundation and helping me become the person I was meant to be. I miss my dad every day, and I'm beyond grateful for the close bond I share with my mom and sister.

To my friends (those who write and those who don't)—we all have a story, and I'm grateful to you for sharing yours with me. Thank you for being in my corner.

To the Orca team—as soon as I saw a book from the Orca Footprints series, I knew I wanted to write one. Thanks for giving me the opportunity and for creating books that get kids interested in the natural world and inspire them to protect it.

YOLANDA RIDGE

is the author of fiction and nonfiction books for young readers, including *Evolution Interrupted: How We Change Nature and How Nature Changes Us*, *CRISPR: A Powerful Way to Change DNA* and *Elliot Jelly-Legs and the Bobblehead Miracle*. She has a master of science degree and is adept at making complex concepts understandable, a skill she uses in her science writing and teaching. Yolanda lives in Rossland, British Columbia.